Ticks Off!
Controlling Ticks That Transmit Lyme Disease on Your Property.

Patrick Guilfoile, Ph.D.

ForSte Press, Inc.
www.forstepress.com

Ticks Off! Controlling Ticks That Transmit Lyme Disease on Your Property. Copyright © 2004 by Patrick Guilfoile. Manufactured in the United States of America. All rights reserved. No part of this book may be reproduced in any form or by any electronic or mechanical means including storage retrieval systems without permission in writing from the publisher, except for brief quotations in reviews of this work. Published by ForSte Press, Inc. P.O. Box 1537, Bemidji, MN 56619. E-mail: books@forstepress.com. Web: www.forstepress.com.

Visit the companion website at www.tickbook.com for links to all the web addresses listed in the book, updates, and other useful information.

ISBN: 0-9753856-0-7

Library of Congress Control Number: 2004092741

Notice:
This book provides information, current as of the date of printing, on strategies that reduce blacklegged tick populations. Every effort has been made to make this manual as accurate and complete as possible. However, differences in geography and ecology between the study sites described in this book and your property may mean some measures will be less effective on your land. Therefore, this text should be used as only a general guide and not as your only source of information about reducing the risk of tick bites and infections associated with tick bites.

Table of Contents

Tick Talk: An Introduction 5

Chapter 1 Personal protection against tick bites 9

Chapter 2 Blacklegged ticks can't jump: The basics of blacklegged tick biology. 17

Chapter 3 Do you have blacklegged ticks (deer ticks) on your property? 23

Chapter 4 Small things first: Deer mice and other hosts. .. 29

Chapter 5 The bigger picture: White-tailed deer 39

Chapter 6 Other strategies for controlling ticks 45

Chapter 7 Blacklegged ticks in the Southeastern U.S. and the Pacific Coast 55

Chapter 8 Putting it all together 61

Literature Cited 63

Glossary 76

Index 78

About the author

Patrick Guilfoile earned his doctorate in Bacteriology at the University of Wisconsin-Madison. He did Postdoctoral work at the Whitehead Institute for Biomedical Research at the Massachusetts Institute of Technology in Cambridge, MA, and has taken additional coursework at The Ohio State University in Acarology (the study of ticks). He is currently a Professor of Biology at Bemidji State University in Bemidji, MN. His research interests include studying several aspects of the biology of ticks and the pathogens they carry. He has co-authored articles on ticks in several scholarly journals including the Journal of Medical Entomology, Journal of Vector Ecology, and Experimental and Applied Acarology.

Acknowledgements

I thank my wife, Audrey, for her assistance and support in nearly every aspect of writing this book. Her work on the cover design and graphic elements added substantially to the visual presentation of this document. The following individuals read the entire manuscript and offered suggestions which greatly improved the book: Dr. Glen Needham, Dr. Derek Webb, Susan Kedzie-Webb, Professor Susan Hauser, David Weld, and Thomas Guilfoile. I would also like to thank the following individuals who read portions of the manuscript and offered valuable, constructive criticism: Dr. James Keirans, Dr. Elizabeth Rave, and Chuck Cole. Janet Crossland at the Peromyscus Genetic Stock Center at the University of South Carolina graciously provided the photograph of *Peromyscus maniculatis*. David Weld of the American Lyme Disease Foundation, Inc. provided the pictures of the 4-poster device for treating deer. Dr. Wayne Rowley of Iowa State University provided recent data on tick distribution in Iowa prior to publication. Dr. Thomas Mather provided information regarding Damminix®. I also thank Art Franz (owner of Photographic Art) for his assistance with several of the illustrations.

Tick Talk: An Introduction

This book is for homeowners, pest management professionals, and others interested in controlling ticks that transmit Lyme disease. Advances in understanding tick biology have recently been translated into improved methods for curbing tick populations. In fact, within the past two years, important developments in tick control have been commercialized, and those innovations are discussed in several places in the book.

This book is designed to assist people who live in areas where ticks transmit Lyme disease and other pathogens. In the U.S., blacklegged ticks (sometimes called deer ticks) transmit the bacterium that currently causes more than 20,000 cases of Lyme disease per year in the United States[1]. Perhaps only $1/3$ to $1/10$ of Lyme disease cases are actually reported[2,3], so the true number of infections is likely to be considerably higher.

Over the past few decades, the range of these ticks has expanded substantially[4,5]. This means significant numbers of people in the U.S., previously protected from tick-borne disease by geography, are unknowingly in the infectious disease crosshairs. Recent news reports highlight this trend. For example:

"Tick-borne diseases at record levels in Minnesota"
Minneapolis Star Tribune (2001)

"Who Let the Ticks Out?"
Medical Society of the State of New York (June, 2003)

"Lyme Disease Spreading"
Wiscasset Newspaper (ME) (June 2003)

"Dreaded Lyme disease bug moves north"
C-Health, 1998 (Canada)

"State sets near-record pace for new Lyme disease cases"
Milwaukee Journal Sentinel (WI), (1999)

"Lyme disease ticks thriving"
Norwich Bulletin (CT), (April, 2002)

"Maryland Lyme disease cases up 36 percent last year"
The Diamondback (U-Maryland), (2000)

This book will help you exercise some control over the daunting expansion of tick-borne disease. I will teach you some specific steps you can take to make your property less hospitable for these parasites. No general guide can provide all the information relevant to your specific area and circumstance. My goal is to provide scientifically supported advice along with recommendations that are reasonable, based on the biology of the tick and the tick's hosts. It should then be possible to apply whatever measures are appropriate for your particular situation.

A brief note on tick names: the ticks now called blacklegged ticks are also frequently referred to as deer ticks. The scientific name for this tick is *Ixodes scapularis*. Until the early-to-mid-1990's, ticks from the Northeastern U.S. were frequently referred to as "*Ixodes dammini*" (or deer ticks) but most tick biologists now consider all these ticks to be the same species, *Ixodes scapularis*. On the west coast of the U.S, a close relative, the western blacklegged tick, *Ixodes pacificus*, transmits Lyme disease as well. In the United States, it is unlikely that you will contract Lyme disease from any other tick species. Although several other *Ixodes* species can transmit Lyme disease, these ticks rarely bite humans. There is also evidence that Lone Star ticks (*Amblyomma americanum*) may transmit a disease that is similar to Lyme disease in parts of the Southeastern United States. The control of the Lone Star tick and the western blacklegged tick are discussed in Chapter 7.

Notes

For your convenience, all of the web addresses in the book are listed as links on the book's companion website: www.tickbook.com. Also note that references, listed by chapter, are in the Literature Cited section in the back of the book.

Tick Facts

The tick's "nose" is located on its front legs (Haller's organ). In addition to smells, this organ detects heat and other cues associated with an approaching animal host.

Blacklegged ticks that carry the Lyme bacterium do not seem to be harmed by the microbe.

White-tailed deer and deer mice are not noticeably harmed by infection with the Lyme bacterium.

Ticks have an incredible arsenal for defense against the host immune response while feeding. The tick's defenses are salivated into the host during feeding and prevent blood clotting, inflammation, activation of immune system cells, and antibody binding. Ticks that can't mount a defense against host immunity will not feed efficiently.

Chapter 1
Personal protection against tick bites

Much of the U.S. population lives in areas where Lyme disease-transmitting ticks are found (Figure 1.1). The risk of contracting Lyme disease from these ticks is particularly high in the coastal areas of the Northeastern U.S. and the Upper Midwest (Figure 1.2). In many of these areas, Lyme disease is commonly acquired in a person's own yard[1, 3].

There is a substantial amount of information already published about personal protection against tick bites (see resources at the end of the chapter) so I won't describe these recommendations in detail. Although a rigorous scientific demonstration of the effectiveness of many of these measures in protecting against Lyme disease is currently lacking[2], these ideas are soundly based on an understanding of the biology of ticks and the Lyme disease bacterium, and are reasonable steps to follow unless and until new information becomes available. A summary of this information is presented below, followed by a list of resources that describe these personal protection measures in more detail.

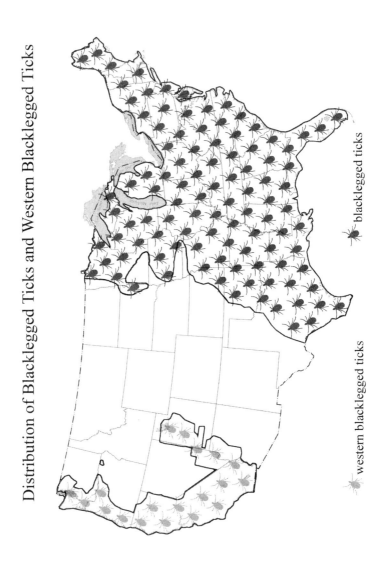

Figure 1.1. Blacklegged tick distribution. Map derived from data from the Centers for Disease Control[21] and other sources[17, 20, 22]. Base map from U.S. Geological Survey (www.nationalatlas.gov).

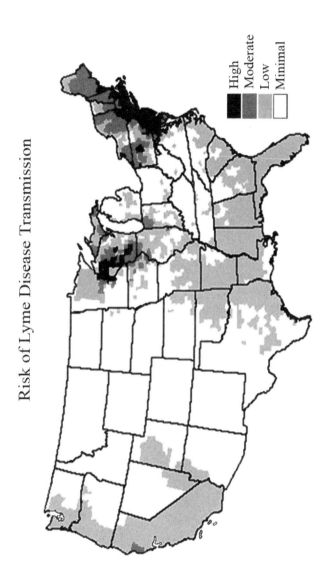

Figure 1.2. Lyme disease risk map. Modified from the Centers for Disease Control[20, 23].

The basic steps in personal protection against tick bites are:
- Avoiding tick-infested areas, if possible.
  - If that is not possible:
- Wear light-colored clothing so that ticks can be easily spotted and removed before they become attached.
- It's not exactly high fashion, but tucking your pant legs into your socks may make it harder for ticks to reach your skin. Many researchers also tape the sock-pants boundary to make it even harder for ticks to find flesh. Anecdotal reports suggest that zip-off pants/shorts (in the "zip-on" state) can trap ticks in the zipper flap on the pants leg, before they reach the waist, which is a common site for tick attachment. (In fact a garment (Tick Chaps™) was developed with just this idea in mind[6].)
- Use repellents and consider the use of pesticides that can be applied to clothing. DEET-based repellents are effective in repelling ticks[7], and The American Academy of Pediatrics states that the use of DEET products with concentrations up to 30%, when appropriately applied, are safe for adults and children over two months of age[14]. Of recently tested DEET-containing products, Ultrathon by 3M had the longest duration of protection against ticks; most other DEET-containing products are considerably less expensive, but provided somewhat shorter protection times[7]. Another repellent, IR3535, is somewhat effective against ticks; plant compounds like citronella do not appear to be very effective[7]. New repellents are being developed and tested, so additional options may be available in the future[19]. Some formulations of permethrin are labeled for application to clothing (prior to the garment being worn). Permethrin kills ticks on contact and will last two weeks or more on

clothing[7]. (As with any pesticide, be sure to follow label directions carefully.) Recent advances in the application of permethrin to clothing may lead to the availability of garments that can be used for years and still retain their effectiveness[13].
- Conduct regular tick checks to ensure that no ticks are biting you. Promptly remove any ticks you find. (Ticks should be stored, as described in Chapter 3, if you plan to have them identified or tested for pathogens.) It is widely accepted that the bacterium that causes Lyme disease is unlikely to be transmitted until ticks have been attached for 24 to 48 hours[11,12], although there may be some circumstances where the Lyme bacterium may be transmitted more rapidly[15]. Regardless, prompt removal of ticks is an important measure in preventing Lyme disease. To remove a tick, use a blunt but narrow tweezers. Grasp the tick as close to the skin as possible, and gently pull the tick straight out of the skin[16,18]. Some commercially-available tick removal tools (Tick plier™, Ticked Off™, and Pro-Tick Remedy™) have been shown to work more effectively than tweezers in removing some species of nymphs[18].
- Heat clothes that may be infested with ticks in the clothes dryer prior to washing. If clothes can't be dried immediately, they should be sealed tightly in a plastic bag to prevent ticks from running loose around your home. A recent report indicates that laundering in a washing machine is not a reliable method for killing ticks on clothing[4]. However, ticks were killed after heating in the dryer for one hour[4]. This suggests that durable, dryer-safe clothes are the appropriate attire if you are walking in potentially tick-infested habitats.

- Lyme disease can occur in dogs and cats, so check pets for ticks and remove them promptly (as described above). Consider treating your pets with an appropriate anti-tick collar, pill, or drop-on medication, as recommended by your veterinarian. A Lyme vaccine is available for dogs.
- Be aware of the signs and symptoms of Lyme disease and other tick-borne diseases (see resources at the end of the chapter). Untreated Lyme disease can be very serious and debilitating. Seek medical attention if you suspect that you may have Lyme disease.

Looking toward the future, another weapon in the personal protection arsenal may include an improved Lyme disease vaccine. A vaccine was commercially available until February, 2002, but is no longer on the market. Efforts continue on the development of an improved vaccine[5] so this may one day be an option for personal protection. In addition, other efforts are underway to develop anti-tick vaccines[9, 10], which would prevent ticks from feeding and potentially transmitting Lyme disease and other infections to humans. An anti-tick vaccine has been developed for cattle in Australia and Cuba, and is also used in parts of South and Central America[8], but much work remains before anti-tick vaccines become a method for preventing Lyme disease. The personal protective measures, described above, are important, but are not the main focus of this book. Rather, the focus is on reducing the blacklegged tick population around your home, as described in the following chapters. Consult the resources listed below for additional information on personal protection against ticks, and for information on the signs and symptoms of Lyme disease and other tick-borne infections.

Other resources on personal protection

Internet resources: (visit our website www.tickbook.com for a complete list of the websites listed in the text, and regular updates as web addresses change):

Centers for Disease Control.
http://www.cdc.gov/ncidod/dvbid/lyme/prevent.htm

State Health Departments. This page, on the CDC website, has links for all the state health departments.
http://www.cdc.gov/other.htm

Mayo Clinic Website
http://www.mayoclinic.com/invoke.cfm?id=DS00116&

American Lyme Disease Foundation, Inc.
http://www.aldf.com

Lyme Disease Foundation.
http://www.lyme.org/ticks/tick.html

Books:
Barbour, A. 1996. *Lyme Disease. The Cause, the Cure, the Controversy.* Johns Hopkins University Press, Baltimore, MD.

Drummond, R. 1998. *Ticks and what you can do about them.* Wilderness Press, Berkley, CA.

Hauser, S. 2001. *Outwitting ticks.* The Lyons Press. New York, NY.

Vanderhoof-Forschner, K. 2003. *Everything you need to know about Lyme disease and other tick-borne disorders.* John Wiley and Sons, Hoboken, N. J.

Tick Facts

The mouthparts of a tick have backward-pointing barbs that allow them to remain firmly attached until they have finished their blood meal. Some ticks (e.g. Lone Star ticks) have long mouthparts that make them especially difficult to remove.

Ticks secrete a protein-rich cement that helps "glue" them in place while they feed. The patch of "skin" that comes off on the tick's mouthparts when you remove a tick is usually this cement.

In some places, so many ticks are present in the environment that even large hosts (like cattle or deer) can die from blood loss due to tick feeding.

Males of some tick species feed near females. Male ticks pump proteins into the host that the female then takes up, allowing her to feed more efficiently.

Chapter 2
Blacklegged ticks can't jump: The basics of blacklegged tick biology.

Blacklegged ticks (also known as deer ticks) have been studied extensively for many years. Consequently, a great deal has been learned about the biology of these ticks, some of which is relevant to efforts to control their populations. One important characteristic of these ticks is their relatively small size, particularly the nymphal ticks (see Figure 2.1, below). This makes it difficult to find and remove these ticks promptly and the longer they feed, the greater the risk they will transmit Lyme disease. (And yes, in case you were wondering about the title of this chapter, blacklegged ticks do not jump on to a host. They wait on vegetation until a host walks by, and then rapidly climb on.)

Figure 2.1. Images of blacklegged (deer) ticks, *Ixodes scapularis*, approximately to scale. From left: Adult female, adult male, nymph, and larva.

Life cycle

Blacklegged ticks are able to survive a variety of extreme weather conditions, and can remain alive for months, waiting for a blood meal[10]. Under favorable conditions, blacklegged ticks can complete their life cycle in as little as two years (egg to egg) (Figure 2.2).

A tick requires a blood meal in order to advance from one life stage to the next. Therefore, these ticks require three blood meals in order to complete their life cycle. In spring, the eggs hatch, releasing larval ticks. Most larval ticks feed on mice or other small rodents, typically during late summer into fall after they hatch (although most of these larvae can make it through the subsequent winter, without a bloodmeal)[10]. Larval ticks that take a blood meal their first summer molt to nymphs in either late summer or the following spring and overwinter in leaf litter or other protected environments. If all goes well for the nymph, it steals a blood meal the next summer, usually from a mouse or other small mammal. The nymph molts to an adult, and the adult seeks a host from late summer into early winter (or the following spring). If the adult female succeeds in getting a meal from a suitable host (usually a white-tailed deer), she will overwinter as an engorged adult. The female will then lay hundreds to thousands of eggs[4] in the spring and the cycle starts over again.

Chapter 2 Blacklegged Tick Biology

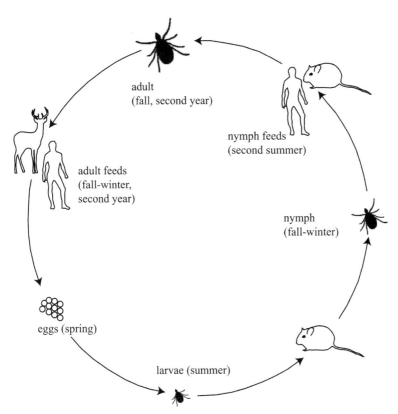

Figure 2.2. Life cycle of the blacklegged tick, with common hosts in the Midwest and Northeast. Note that the life cycle, under the best of conditions (from the tick's perspective) lasts at least two years.

Hosts

Blacklegged ticks feed on a wide variety of animals, ranging from birds to lizards to mammals. These ticks have been reported to feed on over 125 different animals (57 bird, 14 lizard, and 54 mammal species)[1]. Yet just two hosts, deer mice and white-tailed deer, appear to be critical for Lyme disease transmission in the

Northeastern U.S. and upper Midwest[8, 9, 11]. Therefore, much of the book will be devoted to ways to control these animals or the ticks carried by these animals.

Figure 2.3. White-tailed deer, a key host for adult blacklegged ticks, in a typically brushy habitat. Photograph courtesy of Audrey Guilfoile.

The immature stages of blacklegged ticks (larvae and nymphs) typically feed on white-footed or deer mice, which are highly competent as reservoirs for the bacterium that causes Lyme disease[2, 7, 14, 16]. (Competence means that the mice become infected with the Lyme bacterium for a long period of time, and that uninfected ticks can then acquire this pathogen by feeding on an infected animal.) In inland regions of the Northeastern U.S. and in the Midwestern U.S., chipmunks are important hosts for nymphal ticks (and to a lesser extent, larval ticks). Chipmunks are also competent reservoirs for the Lyme bacterium[7,12,13].

Chapter 2   Blacklegged Tick Biology 21

Figure 2.4. The white-footed mouse (shown above) and the deer mouse, which looks very similar, are common hosts for immature blacklegged ticks. Photograph courtesy of the *Peromyscus* Genetic Stock Center and photographer Clint Cook.

Most adult blacklegged ticks feed on white-tailed deer[4,8]. Deer rarely transfer the Lyme bacterium to feeding ticks[7,15], because deer are poor hosts for the Lyme bacterium. Deer are, however, an excellent source of a blood meal for large numbers of ticks. Adult female ticks require a blood meal in order to produce eggs, so deer play an important role in maintaining high population densities of blacklegged ticks.

Blacklegged ticks rarely pass the Lyme bacterium through their eggs to their offspring[5, 6]. Therefore, each generation needs to be re-infected, and this normally happens when blacklegged tick larvae or nymphs feed on deer mice. If the next life stage (a nymph or adult) then latches on to you for a blood meal, you can be infected. Because of their importance in Lyme disease transmission, control of deer mice and deer is covered in separate chapters of the book.

Chapter 3
Do you have blacklegged ticks (deer ticks) on your property?

In some areas of the United States, up to 70% of Lyme disease infections are contracted on a homeowner's property[1, 2, 10]. Therefore, the first step in developing a strategy to "tick-proof" your property is to find out whether blacklegged ticks are present in your area and then on your property. The two best sources of information about whether blacklegged ticks are present in your region are the Centers for Disease control (http://www.cdc.gov/ncidod/dvbid/lyme/index.htm) and your state health department (http://www.cdc.gov/other.htm). Local colleges and universities, extension offices, or county health departments can be contacted for more detailed information on whether blacklegged ticks and Lyme disease are present in your area. Figure 1.1 (Chapter 1) is a useful starting point for determining whether further investigation is needed regarding the presence of blacklegged ticks on your property.

The risk of Lyme disease transmission is not equally high in all areas where blacklegged ticks are present (compare Figures 1.1 and 1.2, Chapter 1). This is the result of many factors including the size of the tick populations, availability of competent hosts, life cycles of ticks in the area, and other variables. The factors responsible for a high proportion of ticks carrying the Lyme bacterium in some areas, but not in others, will be covered in several chapters, particularly in Chapter 7.

As noted previously, in many parts of the country, blacklegged ticks have recently been expanding their range, so these maps only provide a starting point. If you live in or near an area where blacklegged ticks have been reported, the next step is to do a survey on your own property, or hire a lawn care company or pest control company experienced in doing tick surveys.

Probably the most common method for finding ticks is called flagging. You can fashion a flag from a wooden handle or dowel and a white, flannel sheet or white muslin cloth. Specifically, a small flag can be made with a 2 ft. x 3 ft. piece of white cotton muslin or flannel and a wooden stick ~5 ft. long (for example, a broom handle). The narrow side of the fabric is wrapped around the broom handle or dowel and nailed or stapled in place.

The flag is then dragged through vegetation and over the ground, using the stick to stir up the leaf litter, and examined every few feet for the presence of ticks. The darkly colored ticks show up against the white cloth of the flag, and the ticks can be collected from the flag with a tweezers. Once collected, the ticks should be put in a small plastic vial and preserved for identification, as described in more detail later in this chapter.

When using a flag, you want to use the appropriate personal protective measures described in Chapter 1 to reduce your risk of infection if your property is, indeed, infested with blacklegged ticks. (Avoid getting any repellent on the flag, though, or you may be unsuccessful in collecting ticks.)

Checking for blacklegged ticks on animals is another very effective method for determining whether ticks are present on your property. Regularly checking your pets for ticks helps protect your animals from infection, and is an easy way to monitor tick infestation on your property. If you hunt deer, or have someone hunt deer on your property, the deer can be checked for the presence of blacklegged ticks in the fall. Ticks are easiest to find on the head and neck, particularly on and around the ears.

A variety of other methods have been used for tick surveys including the use of carbon-dioxide traps[1, 2, 6]. These traps (which typically use dry ice as a source of carbon dioxide) appear to be fairly ineffective for collecting blacklegged ticks[6,9], compared with flagging.

The timing of surveys to determine whether ticks are present is critical: different life stages of blacklegged ticks are abundant for a relatively short window of time. The general dates listed in Table 3.1 need to be fine-tuned for your location. Typically, nymphs are at peak abundance for only a few weeks in any one place; adults usually have both a fall and spring peak for a few weeks. Your local health department or extension office may be able to provide more specific dates. If not, flagging twice a month during the time spans listed in the table should allow you to effectively sample your property. As a practical matter, adult and nymphal blacklegged ticks are the only stages large enough to be easily collected by a typical homeowner. Different life stages of ticks are normally present in different seasons of the year, and the timing of their appearance depends mainly on the local weather and climate.

Table 3.1: General timing of adult and nymphal blacklegged tick activity in different regions of the country.

Location*	Adult Ticks	Nymphal Ticks
Northeastern U.S.[3]	March-June; Sept. to December	May-September (peak in June- July)
Midwestern U.S.[4]	April- June Sept. to Dec.	May- August
Southeastern U.S.[8]	October-May (peaks in Nov. and Feb.)	April to July (peak from May-June)
Pacific coast of U.S.[5,7]	October-June (peak Jan. through March)	March-August (peak May through July)

Timing will vary somewhat from year-to-year, based on weather and on geographical distance from the locales reported in these articles. *Northeastern U.S. is based on southern New York State; Midwestern U.S. is based on central Minnesota. Southeastern U.S. is based on Southern Missouri. Pacific Coast is based on the San Francisco Bay area. Contact your state health department, local county extension service, or local college or university for more specific dates for tick activity in your locality.

When collecting ticks, preserve them following the guidelines from the organization where you will send or bring ticks for identification. Typically, if you are simply interested in having the tick identified, place it in a small plastic vial with alcohol (rubbing alcohol works fine). It is best not to put the tick on a piece of tape; some features required for identification may be obscured and removing the tick from the tape often damages it to the point where it can't be easily identified. If you are interested in both having the tick identified and having the tick tested for infection with the Lyme bacterium, it is critical to keep the tick alive, since some of these tests can only be performed on live ticks. Keeping the tick in a vial with several blades of grass or a small paper towel or cotton ball moistened with water will keep the tick alive for at least a couple of days. Storage in a standard refrigerator will extend the life of specimens.

If you have a do-it-yourself bent, a very good magnifying glass or a dissecting microscope is required to see the feature necessary to do a preliminary identification. Pressing the tick into a chunk of Blu-Tack® (used for hanging posters) will keep the tick from moving around while you are trying to identify it. For all life stages of ticks that belong to the genus *Ixodes*, the hallmark feature is a pre-anal groove, shown below in Figure 3.3. If you see that feature on a tick, you know it is an *Ixodes* tick.

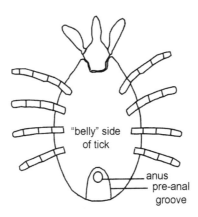

Figure 3.1. Illustration of the pre-anal groove, a diagnostic feature of the Genus *Ixodes*, which includes blacklegged ticks.

Identifying blacklegged ticks to species (e.g. *Ixodes scapularis*) involves determining a number of quite specialized characteristics. Many state or local health departments, extension services, and universities offer tick identification services. Some web references for identification of ticks include:

http://www.entomology.wisc.edu/insectid/index.html
http://www.practicalscience.com/introtick.html
http://www.ksu.edu/parasitology/625tutorials/Tick02.html

(There is a good photograph of the pre-anal groove of *Ixodes* ticks on this last website.)

As of spring, 2004, the following organizations do tick identification (free or for a nominal charge). These web addresses, along with all others in the book, are listed on the book's companion website (www.tickbook.com) for your convenience.

California (Contra Costa Co.) http://www.ccmvcd.dst.ca.us/ticks.htm
Connecticut: http://www.caes.state.ct.us/Tickoffice/tickinfo.htm
Iowa: http://www.ent.iastate.edu/lds/lds.html
Kentucky: http://www.uky.edu/Agriculture/Entomology/entfacts/struct/ef618.htm
Maine: http://www.state.me.us/dhs/boh/ddc/lymeidentification.htm
Massachusetts: http://www.hsph.harvard.edu/now/jun27/bugs.html
Maryland: http://www.agnr.umd.edu/users/hgic/pubs/online/hg501.pdf
Michigan:
http://www.michigan.gov/mda/0,1607,7-125-1566_2403_2421-44271--,00.html
Minnesota (Metro. Area): http://www.mmcd.org/ticks.html
New Jersey: http://www.rce.rutgers.edu/burlington/tick.htm
New York State: http://www.health.state.ny.us/nysdoh/environ/lyme/tickid.htm
New York (Long Island) http://www.lieye.com/articles/ticks/id.shtml
New Hampshire: http://www.ceinfo.unh.edu/common/documents/arthroid.htm
Ohio: http://www.odh.state.oh.us/Resources/publications/IDCManual/dcmweb/Lab29.PDF
Oregon (Jackson Co.) http://www.jacksoncountyvectorcontrol.org/ticklyme.shtml
Rhode Island http://www.uri.edu/news/releases/html/03-0714.html
West Virginia: http://www.wvu.edu/~agexten/ipm/insects/2tick.htm

Chapter 4
Small things first:
Deer mice and other hosts.

As described in Chapter 1, a key aspect of blacklegged tick biology is the hosts from which blood meals are stolen. Ticks require blood to grow, mature, and ultimately produce offspring. Blacklegged ticks normally feed on mice or other small mammals as youngsters (larvae and nymphs), and white-tailed deer as adults. Therefore, an important aspect of blacklegged tick control is to manage mice and deer populations, or, at least, the ticks carried by those mammals.

I focus here primarily on mouse control. As most rural and suburban house dwellers know, controlling mouse populations, even within your home, can sometimes be a challenge. It is therefore unlikely that you could eliminate all mice from your property. In most cases, then, the goal is to keep mouse populations as low as possible.

As a start, eliminating common hiding spots for mice, such as brushpiles, tall grass, and accumulated trash are important. Deer mice are primarily seed eaters, so limiting the availability of unnatural sources of seeds may also be helpful in keeping mouse populations in check. However, you may not need to abandon your bird feeder. On the surface, it would seem that bird feeders could be a potential source of food for deer mice and this, in turn, could lead to increases in mouse and therefore tick populations. Based on a recent investigation, bird feeders do not seem to increase blacklegged tick populations. In this report[7], there was no significant difference in the number of ticks on residential properties with or without bird feeders. In addition, the percentage of ticks with Lyme bacterium (*Borrelia burgdorferi*) infections was similar on properties with and without birdfeeders.

Figure 4.1. A recent report suggests that the use of birdfeeders does not increase the risk of Lyme disease transmission. Photograph courtesy of Audrey Guilfoile.

Chapter 4   Deer Mice and Other Small Mammals   31

This contrasted with two earlier studies, which showed an association of Lyme disease with residences that had bird feeders[8,9]. These earlier reports were case-control studies, meaning that the investigators looked for different characteristics in people who contracted Lyme disease versus those who did not. Such studies, while useful, may obscure a real cause and effect relationship. For example, it is possible that people with birdfeeders also spend more time outdoors, and consequently are more likely to be bitten by Lyme-carrying ticks. Since the more recent study measured tick abundance (and infection of ticks with the Lyme bacterium) on properties with and without bird feeders, I consider it a better indication that bird feeding may not increase risk of Lyme disease transmission, at least in some localities.

Dealing directly with mice and the ticks they carry

Reducing mouse populations directly by trapping and poisoning has not been extensively reported in the scientific literature as a method of reducing the risk of Lyme disease. The difficulty of maintaining low population levels of mice in even a modest-sized yard probably accounts for the scientific silence on this point. One environmentally-friendly alternative is to build or buy nest boxes for owls[20, 21]. These predators may reduce the mouse population and therefore the number of blacklegged ticks in an area. Barn owls, for example, eat up to three deer mice per day[22]. To the best of my knowledge, a link between avian predators of mice and blacklegged tick populations has not been made in the scientific literature. However, this is at least a "feel good" approach, likely helpful for owl populations, and potentially useful in terms of reducing the number of ticks on your property.

A very different approach focuses on the ticks infesting mice, rather than the mice themselves. In one version of this approach, cotton balls treated with a pesticide are distributed over a property. When the mice bring the cotton back to their nests, the insecticide will kill the ticks attached to the mice, or ticks that are loose in the mouse's nest. This product is available commercially under the trade name Damminix®.

Seven studies, all done in the northeastern U.S., found that the numbers of tick larvae (the stage right after hatching from the egg) were reduced in plots treated with pesticide-impregnated cotton[1-6, 10]. Larvae are most likely to feed on mice or other small mammals, so this result was not entirely surprising. Two of these studies found a reduction in blacklegged tick nymphs (the stage most likely to transmit disease to humans) in areas that had been treated with pesticide-impregnated cotton[1, 3]. In one case[1] it was calculated that the risk of Lyme disease transmission was reduced 97% in the treated areas.

Several other studies[4-6], however, did not show a reduction in the number of nymphs in Damminix®-treated areas. The conflicting results in these studies may be the result of at least two factors. One factor may be the size of treated plots. Plot sizes were generally larger in studies that demonstrated a reduction in nymphal ticks. This could the result of nymphs having more difficulty repopulating the central part of the larger treated areas compared to smaller plots used in the other studies.

A second factor may be the degree to which mice were the

Chapter 4 Deer Mice and Other Small Mammals

primary host for immature ticks. In studies where the treatment reduced the number of nymphs, a greater percentage of ticks in those areas may have fed on deer mice, which do use cotton nesting materials. In the studies where no difference was observed, a greater fraction of the larval and nymphal ticks may have fed on other hosts (like chipmunks) that don't use cotton nesting materials.

Homeowners in areas where Damminix® is available (Connecticut, Maryland, Massachusetts, New Hampshire, New Jersey, New York, Pennsylvania, and Rhode Island)[28] may therefore find this product most useful if their property has substantial mouse populations, and if they can treat a relatively large area (an acre or more). This method appears to be quite environmentally benign, with very little of the pesticide (permethrin) spreading from the cotton balls into the soil[10].

Permethrin is considered relatively non-toxic for most mammals, although high concentrations of permethrin have reportedly caused some serious reactions in cats[24]. This might lead to concern about the use of this product in areas where cats may capture or eat treated mice. However, the manufacturer has indicated there had been no reports of adverse reactions in cats in over 15 years of use of the product[28]. Permethrin has low toxicity for birds[26], so it is not likely to have adverse effects on owls and hawks.

According to the manufacturer, the cost of this product is currently about $70 for one application to a typical one-acre lot (with about 50% of that lot being potential mouse habitat). Two applications (in April and July) are normally required[27].

An alternative to the use of cotton nesting material for treating small mammals is the use of baits. These baits are placed in a device that coats or dusts animals with a pesticide as they enter to get the bait. Depending on the baits used, this could potentially target a wider array of potential tick hosts. This approach has been successfully tested for control of another tick (the dog or wood tick)[16]. This method has recently been developed commercially for the control of blacklegged ticks (The MaxForce® Tick Management System by Bayer). In this system, a plastic "maze" is used. The maze contains bait and a wick that doses the mouse with an insecticide (fipronil) as it moves through the maze. According to the manufacturer, following two years of use in an area, this system reduced the number of nymphal ticks in treated plots by 97%[19].

This device was first sold in the spring of 2002, and is expected to be available in most states with a significant Lyme disease risk by the summer of 2004. Fipronil has low toxicity for cats[19] and birds of prey[23, 25], so it should be relatively safe to use in places where cats, hawks, and owls may prey on treated mice. According to the manufacturer, for a typical ~1 acre suburban lot, the cost per application might range from $300 to $700; in most areas, two applications per year would be required[19]. This product is available only through licensed pesticide applicators. Additional information about this device is available from the manufacturer at: www.maxforcetms.com

A variation of the baiting approach is to use food containing a drug that kills ticks as the ticks feed on mice. This method of distributing medicated food has been tested on deer[15] and may

Chapter 4   Deer Mice and Other Small Mammals   35

also be applicable to small mammals. To the best of my knowledge, this latter approach has not yet been developed into a commercial product.

Chipmunks, other small mammals, and birds

As discussed in Chapter 2, chipmunks are important in the Midwest and perhaps other areas as hosts for immature blacklegged ticks. Chipmunks are competent reservoirs for the Lyme bacterium (they can become infected with the Lyme bacterium and can pass this microbe on to uninfected ticks). Therefore, controlling the interactions of ticks with these animals should be considered another important component of a tick-control strategy in many areas.

Figure 4.2. Eastern chipmunks are important host for blacklegged ticks in some areas of the country. Photograph courtesy of Steven Guilfoile.

As noted above, the use of bait boxes containing insecticides is reportedly quite successful for killing ticks on mice. The manufacturer of the MaxForce® Tick Management System) reports this system works for chipmunks as well [19].

Trapping or shooting are other possible methods for controlling chipmunk populations in some rural areas, but these strategies are unlikely to be effective. Chipmunks and most small mammals are territorial, so it is likely that any chipmunks killed on your property would be replaced by migrants from other areas [19].

Maintaining red and gray squirrels populations may be another important element in a Lyme disease control program. These mammals are apparently not readily infected with the Lyme disease bacterium, so ticks feeding on them (instead of mice) are unlikely to acquire and transmit disease [18]. Maintaining or increasing the populations of squirrels may therefore reduce the risk of Lyme disease in an area.

Figure 4.3. A red squirrel, a tick host that may reduce the risk of Lyme disease transmission. Photograph courtesy of Audrey Guilfoile.

Chapter 4 Deer Mice and Other Small Mammals   37

Birds may be important hosts for blacklegged ticks, and at least some birds, such as robins, are capable of being infected with the Lyme bacterium and transmitting this bacterium to feeding ticks[17]. I'm not aware of any published methods for controlling ticks on birds. Since almost all birds are legally protected, the available options are fairly limited.

Looking to the future of tick control involving small animals

A method of tick control that might become an option is to vaccinate mice and other small animals against Lyme disease, perhaps by including a vaccine in food distributed for these animals. In theory, then, vaccinated animals would not be able to support the growth of the Lyme disease pathogen and the incidence of disease should decline. As support of this concept, some lizards are naturally immune to the Lyme bacterium. Ticks feeding on these lizards are cleansed of the Lyme bacterium and this is one factor responsible for the low prevalence of Lyme disease in blacklegged ticks in parts of the United States[14].

This method of vaccinating animals with food baits has been used to reduce rabies prevalence in raccoons and other animals[13]. However, such strategies will probably require years of development if, indeed, they are ever feasible for the control of Lyme disease.

Summary

Immature stages of blacklegged ticks require a bloodmeal, typically from small animals, in order to grow and develop. In many places where Lyme disease is prevalent, deer mice are the preferred hosts for larvae and nymphs. One aspect of controlling tick populations is therefore to control mice (or at least ticks carried on the mice). In addition, control of ticks on other small mammals like chipmunks is also desirable. The recommended control measures include:

- removing hiding places for mice, such as brushpiles
- putting up nest boxes for owls
- removing food sources (don't worry too much about bird feeders, though)
- using pesticide-impregnated cotton (Damminix®) to reduce the population of ticks feeding on mice in your yard (in states where it is available) AND/OR
- using insecticide dousing bait stations (MaxForce® Tick Management System).

Chapter 5
The bigger picture: White-tailed deer

In the Northeastern and Midwestern U.S., the majority of adult blacklegged ticks, in most localities, feed on white-tailed deer[10], and a correlation has been demonstrated between high deer populations and an abundance of blacklegged ticks[8]. Deer are therefore important in establishing and maintaining populations of deer ticks. In many places, large and growing deer populations are associated with newly established or large and growing populations of blacklegged ticks[2-6].

Several approaches to tick control rely on managing deer populations or the ability of deer to carry ticks. Several studies have been published that document the effectiveness of eliminating deer in the control of blacklegged ticks and Lyme disease. In one case[2], killing all deer on an island off the coast of Massachusetts led to a reduction, over several years, of immature blacklegged ticks. Adult ticks looking for a meal, somewhat surprisingly, were easier to collect three years after the deer had been removed. This may be due to an absence of deer as hosts, with the ticks consequently searching harder for a meal from a different host. Eventually (after 8 years) the number of adult blacklegged ticks declined and Lyme disease transmission to humans on that island appears to have ended[11]. Other reports support the notion that almost complete elimination (over 70% or more)[3, 4] of deer is required in order for tick populations to be dramatically reduced. Consequently, in most areas, culling of deer is unlikely to be a useful method of reducing tick populations.

Aside from culling, exclusion of deer using fencing also reduces blacklegged tick populations[7, 9]. An effective fence needs to be at least 8 feet tall and strong enough to prevent entry of deer. A shorter electric fence, in some areas, is another option[9]. The entire lot or yard would need to be fenced (including gates or cattle guards across driveways) making the construction of such a fence costly, and perhaps a violation of local zoning ordinances. Yet, where feasible, such a fence has proven to be very effective in reducing the numbers of ticks in the enclosed areas. For example, in one report[7], the number of nymphal ticks (the life stage most likely to transmit Lyme disease) was reduced by over 80% versus adjacent areas where deer could roam freely. In another report[9], the abundance of nymphs was reduced approximately 50% inside properties that were fenced to exclude deer.

Another approach to tick control relies on treating deer with pesticides, and is based on attracting deer to bait stations. If deer feeding is legal in your locality, a bait station that applies an insecticide as the deer eat is an option to consider. (Feeding deer is controversial[15]. One concern is the potential for increasing deer and therefore blacklegged tick populations. Another concern is disease transmission within herds, and, primarily for this reason, deer feeding is now illegal in a number of places, including New

Chapter 5   Deer

York State[13] and parts of Wisconsin[14].)

Several of these feeding stations have been described in the literature[1,16,17]. One of these, called a "4-poster", has recently been commercially developed, licensed to the American Lyme Disease Foundation, Inc., and produced by Dandux Outdoors[19]. This device is constructed in such a way that deer self-apply a pesticide (permethrin) by touching coated rollers as they feed[1] (Figures 5.1, 5.2). This substantially reduces the amount of pesticide that needs to be applied to an area (as compared with using an insecticide that is sprayed over the ground).

Figure 5.1. 4-poster device for treating deer with an insecticide. Photograph courtesy of the American Lyme Disease Foundation, Inc.

Figure 5.2. Deer feeding at a 4-poster bait station. Photograph courtesy of the American Lyme Disease Foundation, Inc.

This method has been quite effective at reducing tick populations, particularly if used over several years, but these feeding stations are most effective when used over a relatively wide area [1]. In this particular study, four of these deer feeding stations were placed in an area slightly over one-half mile square. Over three years, the investigators found 90% fewer ticks hunting for a meal in the treated plot (compared to an untreated plot) and immature ticks were reduced, on mice, by at least 70%.

The use of these devices requires a serious financial commitment, since each bait station currently costs over $400. In addition, the company selling the feeders intends to sell them in lots of nine, to ensure that deer living in a relatively large area will be treated. There are also operational costs associated with the use of these 4-poster devices, including purchasing feed (up to 1 ton of corn per year, per feeder) and regular pesticide treatments by a licensed applicator. However, with the costs of treating Lyme disease reported to be from a couple of hundred to tens of thousands of dollars per case[12, 18], many homeowners and local units of government may consider this a worthwhile investment.

Long-term studies haven't been completed yet, but it is likely that programs using 4-poster devices for tick control would have to be continued for long periods of time to combat ticks brought into the area by other animals. Of course, local or state regulations will determine whether these deer-baiting stations can be used in your area. The pesticide used in this device can only be applied by a licensed pesticide applicator, and not all states have approved the use of the device (as of Spring, 2004). The 4-poster needs to be placed at least 100 yards from any homes, and signs need to be placed in areas where these devices are used, alerting the public to the presence of the pesticide. In addition, hunters are advised to handle deer with gloves in treatment areas, to avoid contact with the pesticide. In spite of those regulatory requirements, it is likely this device will become an important part of the solution to reducing tick populations in areas where Lyme disease is common.

Summary

The control of deer (or the ticks they carry) can be a critical factor in reducing blacklegged tick populations. Control can take one or more of the following forms, some of which will be more appropriate in particular localities. The control of deer could include:

- Culling of animals to reduce deer populations (this is limited to areas, like islands, where a sustained reduction in deer populations could be maintained).
- Exclusion of deer from your property, using fencing.
- Treating deer with a pesticide (e.g. at a feeding station).

Chapter 6
Other strategies for controlling ticks

At one extreme, a yard completely covered with concrete or asphalt is unlikely to harbor any ticks. In contrast, an overgrown yard with a thick ground cover of leaves, many brush piles, long grass, and abundant food sources for the tick's hosts is likely to provide a highly suitable habitat for large tick populations. In this chapter, I present a balanced view of potential landscape modifications that should reduce tick populations. The goal, in my mind, is to try to find a middle ground of landscape changes that preserve as much of the natural environment of your property as possible, without allowing large tick populations to flourish.

Vegetation modifications
Burning or mowing

Mowing or burning an area should reduce vegetation cover and therefore moisture levels in the environment. Since blacklegged ticks are critically dependent on having access to moist microclimates, burning or mowing has the potential to reduce tick populations.

At least two reports have addressed whether burning (or mowing) can reduce blacklegged tick populations[1,2]. In both cases, the number of ticks was substantially reduced following burning or mowing, with the reduction in tick numbers persisting, in some sites, for up to one year. However, one report[1] suggested that Lyme-infected blacklegged nymphs are more likely to survive a fire, meaning the risk of contracting Lyme disease on a burned tract may not go down in proportion to the reduction in tick numbers. This anomaly is likely the result of more of the surviving ticks having fed on mice, and therefore being more likely to be infected with the Lyme bacterium. Mouse-feeding ticks would be hunkered down in a mouse burrow during a fire. These ticks would be more likely to survive the fire, compared to ticks on the forest floor.

In addition, burned areas often become overgrown with attractive vegetation for deer shortly after a fire[3], which may result in the burned areas rapidly being recolonized with ticks imported by deer. Consequently, burning may not be a long-term solution for reducing tick populations, unless it was repeated frequently or was done in conjunction with other control measures. Burning is tightly regulated in many localities, and proper burning requires more expertise than most homeowners possess. To be effective, the fire must be hot enough to consume almost all the leaf litter and partially decomposed leaves, but not harm larger trees. Controlling a fire so it doesn't damage structures or adjacent property is often a challenge, even for experienced burn crews. Therefore mowing is probably a more realistic option for most property owners.

Neither mowing nor burning completely eliminates ticks. Infected ticks, on a hunt for a blood meal, are frequently found in lawns around homes[4, 5]. In one case, ticks were found (on average) in every square yard of a well-manicured lawn in New York state [4]. Lawns do have a place in a comprehensive plan to manage blacklegged tick populations, however. Although ticks are found

Chapter 6   Other Strategies   47

in lawns, they are much less abundant there, compared to adjacent woodlands. In a study in New York State[5] relatively few adult ticks were found on lawns, although they were abundant in nearby wooded areas. In this study, the more dangerous nymphs were not found on lawns, but were readily collected in nearby woods. In another study, blacklegged nymphs were found on lawns, but the number decreased with increasing distance from a wooded area[6]. Similarly, in Northwestern Wisconsin, ticks were very rarely found on lawns of farmsteads, although they were present in substantial numbers in adjoining woodlots[22]. Therefore mowed areas can offer some, albeit incomplete, protection from blacklegged ticks.

Cutting down trees and shrubs

The previously mentioned reports indicating ticks are less abundant on lawns suggest that removal of trees and shrubs can reduce tick populations. For Lone Star ticks, thinning forests does reduce tick populations[27], and this may apply to blacklegged ticks as well. Some limited tree and shrub removal would be expected to reduce moisture levels near the ground and may therefore be a reasonable part of a program to reduce the number of blacklegged ticks on your property. Blacklegged ticks avoid direct sunlight, so creating openings in the forest canopy should reduce the amount of suitable tick habitat.

Vegetation plantings

There has been only limited investigation of plants that inhibit tick colonization of an area. For example, the planting of certain grasses that exude sticky or noxious substances can interfere with tick questing[reviewed in 19]. In addition, the planting of evergreens, which do not produce extensive amounts of leaf litter and tend to acidify the soil, may also be a useful long-term strategy for tick control. In one study conducted in Wisconsin and Illinois, blacklegged ticks were less likely to be found in coniferous forests, compared to deciduous forests[21]. In another report[22] ticks on farms were much more abundant in deciduous woodlands, as compared with areas under and adjacent to evergreen (spruce or pine) shelterbelt trees. Yet, hemlock forests, at least in Pennsylvania, appear capable of sustaining blacklegged ticks[20]. Coniferous forests are therefore not universally unsuitable tick habitats.

Leaves

As tough as ticks are, their Achilles heel (if they had heels) is a lack of moisture. If kept in a dry environment, a tick will die. Therefore, one critical set of landscape changes is to remove moist microenvironments from your property. This includes leaves and leaf piles, which can serve as a year-round source of moisture and provide insulation that may allow ticks to survive cold temperatures that would otherwise kill them.

The concept of leaf removal to reduce tick populations was investigated in a wooded area, surrounded by residential lots, in New Jersey[7]. Plots about twice the size of a football field were compared, some with leaf litter removed, others with the leaves left undisturbed. Leaves were removed from the plots in either March or June. About 75% fewer nymphs were present in plots with no leaves, compared to those with leaves in place. Leaf removal is therefore on option that should be considered in a tick control program, particularly in areas near a residence. Based on this report, though, leaf removal needs to be very thorough, with both leaves and the duff layer (partially decomposed leaves) needing to be removed. This means that leaf blowers are probably not sufficient for leaf removal, unless the area has been cleared of leaves each year.

Non-vegetation-related control measures
Tick traps, flagging to collect ticks

The use of CO_2 traps has been advocated to reduce tick populations in residential areas[18]. At least some species of ticks are attracted to CO_2 given off by animals, so there is a reasonable biological basis to this idea. These traps typically consist of a block of dry ice surrounded by tape designed to trap ticks that crawl toward the source of carbon dioxide. These traps have been used to sample tick populations in various studies[e.g. 4, 5]. However, published reports suggest that these traps are unlikely to substantially reduce blacklegged tick populations over large areas. As one example[5], when CO_2 traps were used to sample nymphs, only about 20% of the nymphs were trapped in the tape surrounding the traps; most of the remainder were still on the hunt for a meal in the vicinity of the trap. In addition, adult blacklegged ticks travel an average of less than a foot a day[25], so the number of traps required to treat a large area would be prohibitive.

Flagging is useful for determining whether ticks are present on your property, but is unlikely to reduce tick populations substantially. In one study, about 7% of blacklegged nymphs in an area were picked up by flagging[29], not an encouraging statistic in terms of reducing tick populations.

Pesticides

Many people would prefer to avoid the use of pesticides, if possible[16]. Yet a substantial body of research indicates pesticides are very effective in reducing tick populations[8-15]. Before I go into detail about anti-tick pesticides, please be aware that regulations concerning pesticides frequently change, and it is advisable to check with your local university extension agent, or other professional regarding the pesticide options available for killing ticks in your locality.

In many cases, the most effective pesticides can only be used by a licensed pesticide applicator. Also, the targeted application of pesticides to the tick's hosts is probably the first strategy that should be attempted, as described in Chapters 4 and 5. This greatly reduces the amount of pesticide that needs to be applied to a given area, and reduces any attendant environmental risks. However, there may be some situations where additional pesticide applications may be needed to effectively control tick populations. If you are applying the chemicals yourself, be sure to read pesticide labels completely before using these chemicals. If there is a contradiction between the information given here and the label directions, be sure to follow the instructions on the label.

For a homeowner, one approach to chemically-based tick control is to use non-residual pesticides (e.g. insecticidal soap and permethrin formulations). Because they break down quickly, these pesticides are considered more environmentally benign, as compared with most other pesticides. This advantage is also a liability, though, in that the period of tick control with these chemicals probably doesn't exceed two weeks[8, 9]. These treatments may have a place in a regime where they are applied several times during the peak of blacklegged nymph abundance, generally in early to mid-summer. A number of pesticides that are chemically related to permethrin, such as deltamethrin (Suspend®, Deltaguard G) are also useful for controlling blacklegged ticks[32].

Chapter 6 Other Strategies

Many of these pesticides are available only to licensed pesticide applicators. Several residual pesticides are effective for long-term control of blacklegged ticks[10-15]. Two of these pesticides have recently been taken off the market, however. Registration for home use of chloropyrifos (Dursban) expired in 2001[17] and diazinon will no longer be available for outdoor use by homeowners by the end of 2004[23]. Carbaryl (Sevin)[10, 11, 13, 14, 15] and cyfluthrin[10, 12] are two residual pesticides that are still available for tick control in many areas. Uses of carbaryl are being restricted, however[32] and cyfluthrin is available only to licensed pesticide applicators in many states. With these pesticides, control of blacklegged ticks is frequently 90% or higher, and the population reductions usually persist for months. See table 6-1 below.

Table 6-1. Control of Blacklegged Ticks with Carbaryl and Cyfluthrin

Pesticide	Tick Reduction*	Length of the Study	References
Carbaryl	72%	3 months	15
	68-78%	41 days	10
	70-90%	~1 month	11
	87-97%	5 months	14
	60-100%	2 months	12
	90-100%	~5 months	13
Cyfluthrin	87-100%	41 days	10
	100%	1 year	12

*Different studies used different amounts of pesticide and frequently determined tick reduction differently. Therefore, the studies are not strictly comparable with one another.

There are other considerations regarding the use of these pesticides besides effectiveness against ticks. With some pesticides, acute human toxicity from overexposures may be a concern. The effects of these chemicals on non-target insects and other animals are also factors that need to be considered. In your locality, the services of a licensed pesticide applicator would probably be needed in order to use at least some formulations of these chemicals. If you are able to purchase an appropriate product that is available to homeowners, be sure to carefully follow label directions. The National Institutes of Health has recently made a database available with details on chemicals in household products including pesticides[30]. It contains information on the chemical content of specific products, along with potential health and environmental effects of these chemicals, and can give you ideas for appropriate products, before you visit a store.

Biological Control

Currently, options for biological control of blacklegged ticks are limited, but, in the future, more options may become available. For example fungi, wasps, and other organisms that kill blacklegged ticks have been identified[reviewed in 19]. None of these parasites are commercially available, and there is some evidence that, at least in the case of wasps, very high populations of blacklegged ticks are required to support substantial wasp populations[26]. (These wasps only attack ticks and similar creatures, so there isn't a concern about people being stung by them.)

Since parasite populations are normally regulated in such a way that only a fraction of the host population is killed, laboratory-reared fungi, wasps, and other biological control agents would probably have to be introduced in large numbers to an area, in order to effectively control blacklegged ticks[28]. It is not likely that blacklegged tick populations would be eliminated by these biological control methods, but they may become an important part

Chapter 6   Other Strategies

of an integrated program to reduce blacklegged tick populations.

Healthy natural environments

Maintaining a diverse, healthy woodland habitat is another potential way to decrease blacklegged tick populations on your property. Habitats with larger populations of squirrels and other mammals that don't become infected with the Lyme bacterium, and smaller populations of deer mice will likely result in less transmission of Lyme disease to humans[24]. Ticks that feed on squirrels, skunks, lizards, and other animals are less likely to be infected with the bacterium that transmits Lyme disease, compared to ticks that feed on deer mice or chipmunks[24]. The best method for ensuring a diverse set of mammal species in your area is to maintain large, unbroken expanses of woodland areas. This may be a useful argument for setting aside additional land, in suburban or rural areas, as forest preserves.

Buffers of unsuitable habitat

Physically separating your lawn from woods and other areas with high tick populations has been reported to reduce the number of ticks on adjacent lawns by 50%[31]. This method involves placing a dry barrier (for example, gravel or pumice) as a buffer between lawns and other frequently used areas, and adjacent woodlands. This buffer makes it more difficult for ticks to crawl onto a lawn or other area frequented by people. It doesn't form an absolute barrier, though, since birds and mammals, which play an important role in tick dispersal, can easily cross this strip. Any such barrier would probably work best if it was several feet wide, and was regularly maintained so leaf litter and other materials that retained moisture didn't accumulate.

Summary

Based on the information presented in this chapter, the primary recommendations for landscaping for tick control include:
- Mowing parts of your property that regularly have human traffic or are adjacent to areas with human traffic.
- Removing leaves and leaf litter every fall or spring.
- Judicious use of pesticides.
- Trying to maintain tracts of undeveloped, healthy ecosystems in your local area.
- Creating barriers to tick movement between wooded areas and areas frequented by people.

Chapter 7
Blacklegged ticks in the Southeastern U.S. and the Pacific Coast

Most of the information in the previous chapters applies to ticks that carry the Lyme bacterium, regardless of geographical location. Yet the ecology of some aspects of Lyme disease differs along the Pacific Coast and Southeastern U.S. in several critical respects.

In some areas of the Southeastern U.S. and the West Coast, blacklegged ticks are abundant, but Lyme disease is less common than in the Northeastern U.S. and Midwestern U.S. One reason is that relatively few ticks in these areas are infected with the Lyme bacterium. For example, the incidence of infected western blacklegged ticks is typically 1-2% along the Pacific Coast[5, 13] with some pockets of higher prevalence[14]. In the Southeastern U.S., an estimate of 2%-8% of Lyme bacterium-infected ticks has been published[12]. This prevalence of infected ticks is much lower than along the East Coast (frequently 50% or higher in adult ticks in many of these areas)[5] and the Midwest (in the teens or higher)[16].

At least some of this difference in Lyme disease incidence is probably the result of different tick hosts in these areas. Some of the blacklegged tick hosts in the Southeast U.S. and Pacific Coast are poor reservoirs of the Lyme disease bacterium, or actually eliminate the Lyme bacterium from ticks that feed on them. For example, in these areas of relatively low Lyme disease incidence, immature ticks frequently feed on lizards. Lizard blood, in some cases, can kill the Lyme bacterium in the gut of the tick, preventing further transmission[6,7]. Therefore, Lyme disease control measures in these areas could include interventions that maintain or increase lizard populations (although some lizards are apparently competent hosts for the Lyme bacterium)[12].

In California and perhaps other areas on the Pacific Coast, Lyme disease transmission potentially involves one or more small mammals (dusky-footed woodrats, California kangaroo rat, and three species of deer mice)[8], all of which are competent reservoirs for the Lyme bacterium. Ticks were controlled on Mexican woodrats in Colorado using bait tubes containing an insecticide[1]. Therefore, baiting methods (like the MaxForce® Tick Management System) may also have a place in tick control along the Pacific Coast. Columbian black-tailed deer appear to be important hosts for the adult stages of the western blacklegged tick[5], so the application of control measures for deer, detailed in Chapter 5, may be important as well.

One other potential difference between the western blacklegged tick and the blacklegged tick may be important in Lyme disease transmission. Relatively few blacklegged ticks are infected with the Lyme bacterium at the time they hatch from the egg, and relatively few, which are infected, have the Lyme bacterium outside their gut[11]. In contrast, western blacklegged ticks more readily transmit the Lyme bacterium from infected female to offspring, suggesting that infection of these ticks with the Lyme bacterium may persist without each generation of ticks needing to be re-infected. As noted above, fewer western blacklegged ticks are typically infected with the Lyme bacterium, compared with blacklegged ticks in the Midwest and Northeastern U.S. However, more of the infected western blacklegged ticks have the Lyme bacterium spread throughout their entire bodies, potentially leading to more rapid transmission of the Lyme bacterium to a human[5]. This possibility suggests that a strategy of frequent tick checks may be even more important in areas where western blacklegged ticks are common.

In the Southeastern U.S., several small mammals appear to be important in maintaining the Lyme bacterium in nature. These include the cotton mouse, cotton rat, eastern woodrat, and cottontail rabbit. These mammals can be naturally infected with the Lyme bacterium and may be important reservoirs for Lyme transmission[12]. The cycles involving Lyme transmission among mammals and ticks in the Southeastern U.S. appear to be more complicated than those involving primarily deer mice and white-tailed deer in the Midwest and Northeast. These complications may help lower the risk of infection in the Southeast, since some of the potential tick hosts (e.g. at least some species of lizards) are likely incapable of transferring the Lyme bacterium to ticks[12].

Because of the complexity of the cycles involving Lyme disease transmission, and a lack of clear data on how frequently humans become infected with the Lyme bacterium in the Southeastern U.S., it is difficult to suggest clear-cut recommendations for managing properties in the Southeastern U.S. to minimize blacklegged tick populations. The personal protective measures described in Chapter 1 apply, and hopefully, as more information becomes available, more satisfactory recommendations can be made.

Lone Star ticks and STARI

An additional complication involving Lyme or Lyme-like disease in the Southeastern U.S. is that Lone Star ticks can carry a microbe similar to the Lyme disease bacterium. This bacterium is thought to cause a disease caused Southern Tick-Associated Rash Illness (STARI), also sometimes called Master's disease or Southern Lyme disease[17]. Investigations are currently underway to better understand the degree to which this microorganism (*Borrelia lonestari*) causes disease in humans. This bacterium has recently been cultured[19], and this should result in a better understanding of this pathogen and the disease it causes. This microbe appears to be widespread; it has been found in Lone Star ticks across the southern and eastern United States[3]. The ticks themselves are primarily found in the Southeastern and East-Central United States, and along the Eastern Seaboard (Figure 7.1).

Chapter 7 Southeast U.S. and Pacific Coast

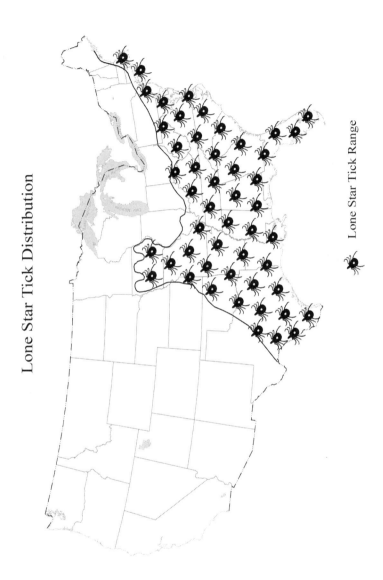

Figure 7.1. Distribution of the Lone Star tick in the United States. Data from the Centers for Disease Control and other sources[9, 10, 18]. Base map from U.S. Geological Survey (www.nationalatlas.gov).

In terms of control of Lone Star ticks, most of the methods described in previous chapters apply. There is evidence that control of deer by fencing[2], vegetation control to lower humidity at ground level[4] and use of pesticides is effective at reducing the number of Lone Star ticks[4]. All of these methods of tick control have been described previously in this book.

There are several differences in the biology of blacklegged ticks and Lone Star ticks, which may be important for controlling populations of Lone Star ticks. Lone Star ticks tend to be more aggressive in looking for hosts and travel longer distances in search of a host[10]. Consequently, carbon dioxide traps are fairly effective in attracting Lone Star ticks. These traps may therefore be a useful method for surveying your property for the presence of these ticks. In addition, Lone Star ticks are larger, and therefore are easier to see and remove before they have a chance to attach, compared to blacklegged ticks.

Chapter 8
Putting it all together

Integrated Pest Management is an increasingly important concept for dealing with a variety of harmful insects and other vermin. The basic tenet of Integrated Pest Management is that multiple methods of control, applied simultaneously, are often more effective than a single measure. To date, almost all studies concerning blacklegged tick control have focused on a single approach to reducing tick populations. The effectiveness of simultaneously using multiple measures to control blacklegged ticks is likely to be greater than using a single method, but this hypothesis has yet to be critically tested. Regardless of the methods used, controlling tick populations requires patience. With the exception of area spraying of insecticides, most of the measures described in the book will require months or even years before they are fully effective.

In the case of the control of blacklegged ticks, it is important to realize that no measures are likely to completely eliminate all ticks. Consequently, the personal protective measures described in the first chapter should be used in conjunction with these other approaches.

It is also important to realize that reducing the ability of the primary tick hosts— deer mice and deer— to feed ticks is probably the most critical step in controlling tick populations.

The final thing to keep in mind is that some vegetation changes, principally those that make it harder for ticks to conserve moisture, should also reduce tick numbers. Based on those principles, here are my recommendations:

- Establish whether blacklegged ticks are present in your area and on your property (Chap. 3). If they are:
- Reduce numbers of mice by eliminating sources of food and shelter; consider installing owl nest boxes (Chap. 4)
- Put out insecticide-treated cotton (Damminix®) on your property, if this product is available in your state (Chap. 4)
- Consider hiring a professional to place bait boxes which coat mice with pesticides as they navigate the maze in search of bait (MaxForce® Tick Management System)(Chap. 4)
- Consider measures to control chipmunk populations and work to maintain populations of squirrels (Chap. 4).
- Consider erecting fencing to exclude deer from your property (Chap. 5)
- If fencing is not an option, consider cooperating with your neighbors to install deer feeders that will treat deer with pesticide as they eat (Chap. 5)
- Consider mowing an area around your house (Chap. 6)
- Remove leaf litter from lawns and from wooded areas adjacent to lawns (Chap. 6)
- Consider the judicious use of pesticides to further control tick populations, particularly along trails and in wooded areas adjacent to lawns and other traffic areas (Chap. 6)
- Create a barrier of gravel or similar material that will make it difficult for ticks to migrate from forested areas into areas frequented by people (Chap. 6).

Literature Cited

References for the Introduction

1) Centers for Disease Control. 2003. Notice to Readers: Final 2002 Reports of Notifiable Diseases. Morbidity and Mortality Weekly Report. 52(31);741-750

2) Naleway A., E. Belongia, J. Kazmierczak, R. Greenlee, and J. Davis. 2002. Lyme disease incidence in Wisconsin: A comparison of state-reported rates and rates from a population-based cohort. American Journal of Epidemiology 155(12):1120-1127

3) Meek J, C. Roberts, E. Smith Jr, M. Cartter. 1996. Underreporting of Lyme disease by Connecticut physicians, 1992. Journal of Public Health Management Practices 2(4):61-65.

4) Sanders, K. and P. Guilfoile. 2000. New records of the blacklegged tick, *Ixodes scapularis* (Acari:Ixodidae) in Minnesota. Journal of Vector Ecology 25(2): 155-157.

5) White, D., H. Chang, J. Benach, E. Bosler, S. Meldrum, R. Means, J. Debbie, G. Birkhead, and D. Morse. 1991. The geographic spread and temporal increase of the Lyme disease epidemic. JAMA 266(9): 1230-1236.

References for Chapter 1

1) Falco, R. and D. Fish. 1988. Prevalence of *Ixodes dammini* near the homes of Lyme disease patients in Westchester County, N.Y. American Journal of Public Health. 127:826-830.

2) Poland, G. 2001. Prevention of Lyme disease: A review of the evidence. Mayo Clinic Proceedings 76:713-724

3) Falco, R. and D. Fish. 1988. Ticks parasitizing humans in a Lyme disease endemic area of southern New York state. American Journal of Epidemiology 128(5): 1146-1152.

4) Carroll J. 2003. A cautionary note: survival of nymphs of two species of ticks (Acari: Ixodidae) among clothes laundered in an automatic washer. Journal of Medical Entomology 40(5): 732-736.

5) Willett, T., A. Meyer, E. Brown, and B. Huber. 2004. An effective second-generation outer surface protein A-derived Lyme vaccine that eliminates a potentially autoreactive T cell epitope. Proceedings of the National Academy of Sciences, USA 101:1303-1308.

6) Heitzel, J.A. Learning about Lyme Disease. 3M Corporation p25

7) Consumer Reports. Safe use of Repellents. The Buzz on Repellents. May, 2003, p 14-15

8) Fuente, J., M. Rodriguez, C. Montero, M. Redondo, J. Garcia-Garcia, L. Mendez, E. Serrano, M. Valdes, A. Enriquez, M. Canales, E. Ramos, O. Boue, H. Machado, R. Lleonart. 1999. Vaccination against ticks (*Boophilus* spp.): the experience with the Bm86-based vaccine Gavac™. Genetic Analysis: Biomolecular Engineering 15: 143-148.

9) Almazan, C., K. Kocan, D. Bergman, J. Garcia-Garcia, E. Blouin, and J. Fuente. 2003. Identification of protective antigens for the control of *Ixodes scapularis* infestations using cDNA expression library immunization. Vaccine 21:1492-1501.

10) Valenzuela, J. I. Franscishetti, V. Pham, M. Garfield, T. Mather, and J. Riberio. 2002. Exploring the sialome of the tick *Ixodes scapularis*. The Journal of Experimental Biology. 205: 2843-2864.

11) de Silva, A. and E. Fikrig. 1995. Growth and migration of *Borrelia burgdorferi* in *Ixodes* ticks during blood feeding. American Journal of Tropical Medicine and Hygiene. 53(4): 397-404.

12) Piesman, J. G. Maupin, E. Campos, and C. Happ. 1991. Duration of adult female *Ixodes dammini* attachment and transmission of *Borrelia burgdorferi*, with description of a needle aspiration isolation method. Journal of Infectious Diseases 163:895-897.

13) Faulde MK, W. Uedelhoven and R Robbins. 2003. Contact toxicity and residual activity of different permethrin-based fabric impregnation methods for *Aedes aegypti* (Diptera: Culicidae), *Ixodes ricinus* (Acari: Ixodidae), and *Lepisma saccharina* (Thysanura: Lepismatidae). Journal of Medical Entomology 40(6):935-41

14) American Academy of Pediatrics. West Nile Virus Information. http://www.aap.org/family/wnv-jun03.htm

15) Shih, C-M. and A. Spielman. 1993. Accelerated transmission of Lyme disease spirochetes by partially fed vector ticks. Journal of Clinical Microbiology 31(11): 2878-2881.

16) Needham, G. 1985. Evaluation of five popular methods for tick removal. Pediatrics 75: 997-1002.

17) Dr. Wayne Rowley, Iowa State University, personal communication.

18) Stewart, J. R., W. Burgdorfer, and G. Needham. 1998. Evaluation of three commercial tick removal tools. Wilderness and Environmental Medicine 9:137-142.

Literature Cited 65

19) Carroll, J., V. Solberg, J. Klun, M. Kramer, and M. Debboun. 2004. Comparative activity of Deet and AI3-37220 repellents against the ticks *Ixodes scapularis* and *Amblyomma americanum* (Acari: Ixodidae) in laboratory bioassays. Journal of Medical Entomology 41(2): 249-254.

20) Sanders, K. and P. Guilfoile. 2000. New records of the blacklegged tick, *Ixodes scapularis* (Acari:Ixodidae) in Minnesota. Journal of Vector Ecology 25(2): 155-157.

21) Centers for Disease Control. Map of reported and established *Ixodes scapularis* populations. http://www.cdc.gov/ncidod/dvbid/lyme/tickmap.htm

22) Drummond, R. 2000. Ticks and what you can do about them. p 24. Wilderness Press, Berkley, CA.

23) Centers for Disease Control Lyme disease risk map. http://www.cdc.gov/ncidod/dvbid/lyme/riskmap.htm

References for Chapter 2

1) Keirans, J., H. Hutchinson, L. Durden, and J. Klompen. 1996. *Ixodes (Ixodes) scapularis* (Acari: Ixodidae): Redescription of all active stages, distribution, hosts, geographical variation, and medical and veterinary importance. Journal of Medical Entomology. 33(3): 297-318.

2) Donahue, J., J. Piesman, and A. Spielman. 1987. Reservoir competence of white-footed mice for Lyme disease spirochetes. American Journal of Tropical Medicine and Hygiene 36: 92-96.

3) Fish, D. 1995. Environmental risk and prevention of Lyme disease. The American Journal of Medicine. 98(Suppl. 4A): 2S-9S.

4) Wilson, M., T. Litwin, T. Gavin, M. Capkanis, D. Maclean, and A. Spielman. 1990. Host-dependent differences in feeding and reproduction of *Ixodes dammini* (Acari: Ixodidae). Journal of Medical Entomology 27(6): 945-954.

5) Magnarelli, L., J. Anderson, and D. Fish. 1987. Transovarial transmission of *Borrelia burgdorferi* in *Ixodes dammini*. Journal of Infectious Diseases 156:234- 236

6) Piesman, J., J. Donahue, T. Mather, and A. Spielman. 1986. Transovarially acquired Lyme disease spirochetes (*Borrelia burgdorferi*) in field-collected larval *Ixodes dammini* (Acari:Ixodidae). Journal of Medical Entomology 23:219-220.

7) LoGiudice, K., R. Ostfeld, K. Schmidt, and F. Keesing. 2003. The ecology of infectious disease: Effects of host diversity and community composition on Lyme disease risk. Proceedings of the National Academy of Sciences, USA. 100(2):567-571.

8) Wilson, M., G. Adler, and A. Spielman. 1985. Correlation between abundance of deer and that of the deer tick, *Ixodes dammini* (Acari:Ixodidae). Annals of the Entomological Society of America 78: 172-176.

9) Jones, C., R. Ostfeld, M. Richard, E. Schauber, and J. Wolff. 1998. Chain reactions linking acorns to Gypsy Moth outbreaks and Lyme disease risk. Science 279: 1023-1026.

10) Yuval, B. and A. Spielman. 1990. Duration and regulation of the development cycle of *Ixodes dammini* (Acari: Ixodidae). Journal of Medical Entomology 27:196-201.

11) Lane, R., J. Piesman, and W. Burgdorfer. 1991. Lyme borreliosis: Relation of its causative agent to its vectors and hosts in North America and Europe. Annual Review of Entomology 36:587-609.

12) Slajchert, T., U. Kitron, C. Jones, and A. Manelli. 1997. Role of the eastern chipmunk, (*Tamias striatus*) in the epizootiology of Lyme borreliosis in Northwestern Illinois, USA. Journal of Wildlife Diseases 33(1): 40-46.

13) Schmidt, K., R. Ostfeld, and E. Schauber. 1999. Infestation of *Peromyscus leucopus* and *Tamias striatus* by *Ixodes scapularis* (Acari:Ixodidae) in relation to the abundance of hosts and parasites. Journal of Medical Entomology 36(6): 749-757.

14) Levine, J., M. Wilson, and A. Spielman. 1985. Mice as reservoirs of the Lyme disease spirochete. American Journal of Tropical Medicine and Hygiene 34: 355-360.

15) Telford, S. III, T. Mather, S. Moore, M. Wilson, and A. Spielman. 1988. Incompetence of deer as reservoirs of the Lyme disease spirochete. American Journal of Tropical Medicine and Hygiene 39:105-109.

16) Mather, T., S. Telford III, S. Moore, and A. Spielman. 1990. *Borrelia burgdorferi* and *Babesia microti*: Efficiency of transmission from reservoirs to vector ticks (*Ixodes dammini*). Experimental Parasitology 70: 55-61.

References for Chapter 3

1) Falco, R. and D. Fish. 1988. Prevalence of *Ixodes dammini* near the homes of Lyme disease patients in Westchester County, N.Y. American Journal of Public Health. 127:826-830.

2) Maupin, G., D. Fish, J. Zultowsky, E. Campos and J. Piesman. 1991. Landscape ecology of Lyme disease in a residential area of Westchester County, N.Y. American Journal of Epidemiology 133:1105-1113.

3) Fish, D. 1995. Environmental risk and prevention of Lyme disease. The American Journal of Medicine. 98(Suppl. 4A): 2S-9S.

Literature Cited

4) Minnesota Department of Health Website. http://www.health.state.mn.us/divs/idepc/diseases/lyme/ldsldtxt.html

5) Li, X., C. Peavey, and R. Lane. 2000. Density and spatial distribution of *Ixodes pacificus* (Acari: Ixodidae) in two recreational areas in north coastal California. American Journal of Tropical Medicine and Hygiene 62(3): 415-422.

6) Ginsberg, H. and C. Ewing. 1989. Comparison of flagging, walking, trapping, and collecting from hosts as sampling methods for northern deer ticks, *Ixodes dammini*, and lone-star ticks, *Amblyomma americanum* (Acari: Ixodidae). Experimental and Applied Acarology 7: 313-332.

7) Padgett, K., and R. Lane. 2001. Life Cycle of *Ixodes pacificus* (Acari: Ixodidae): Timing of Developmental Processes Under Field and Laboratory Conditions. Journal of Medical Entomology 38: 684-693.

8) Kollars TM Jr, J. Oliver Jr, P. Kollars, L. Durden. 1999. Seasonal activity and host associations of *Ixodes scapularis* (Acari: Ixodidae) in southeastern Missouri. Journal of Medical Entomology 36(6):720-726.

9) Falco, R. and D. Fish 1991. Horizontal movement of adult *Ixodes dammini* (Acari: Ixodidae) attracted to CO_2-baited traps. Journal of Medical Entomology 28: 726-729.

10) Falco, R. and D. Fish. 1988. Ticks parasitizing humans in a Lyme disease endemic area of southern New York state. American Journal of Epidemiology 128(5): 1146-1152.

References for Chapter 4

1) Mather, T., J. Ribeiro, S. Moore, and A. Spielman. 1988. Reducing the transmission of Lyme disease spirochetes in a suburban setting. Annals of the New York Academy of Sciences. 539: 402-403.

2) Mather, T., J. Ribeiro, and A. Spielman. 1987. Lyme disease and Babesiosis: Acaracide focused on potentially infected ticks. American Journal of Tropical Medicine and Hygiene 36: 609-614.

3) Deblinger, R., and D. Rimmer. 1991. Efficacy of a permethrin-based acaricide to reduce the abundance of *Ixodes dammini* (Acari: Ixodidae). Journal of Medical Entomology 28: 708-711.

4) Daniels, T., D. Fish, and R. Falco. 1991. Evaluation of host-targeted acaricide for reducing risk of Lyme disease in southern New York State. Journal of Medical Entomology 28:537-543.

5) Stafford, K. 1991. Effectiveness of host-targeted permethrin in the control of *Ixodes dammini* (Acari: Ixodidae). Journal of Medical Entomology. 28: 611-617.

6) Stafford, K. 1992. Third-year evaluation of host-targeted permethrin for the control of *Ixodes dammini* (Acari: Ixodidae) in southeastern Connecticut. Journal of Medical Entomology. 29: 717-720.

7) Townsend, A., R. Ostfeld, and K. Geher. 2003. The effects of bird feeders on Lyme disease prevalence and density of *Ixodes scapularis* (Acari: Ixodidae) in a residential area of Duchess County, New York. Journal of Medical Entomology. 40: 540-546.

8) Orloski, K., G. Campbell, C. Genese, J. Beckley, M. Schriefer, K. Spitalny, and D. Dennis. 1998. Emergence of Lyme disease in Hunterdon County, New Jersey, 1993: a case-control study of risk factors and evaluation of reporting patterns. American Journal of Epidemiology 147:391-397.

9) Smith, G., E. Wileyto, R. Hopkins, B. Cherry, and J. Maher. 2001. Risk factors for Lyme disease in Chester County, PA. Public Health Reports Supplement 1, 116: 46-156.

10) Ginsberg, H. 1992. Managing ticks and Lyme spirochetes: Efficacy and Potential Environmental Impact of Permethrin-treated cotton balls. In: Ecology and Management of Ticks and Lyme disease at Fire Island National Seashore and selected Eastern National Parks. NPS/NRSUNJ/NRSM-92/20 U.S. Dept. of Interior, National Park Service

11) Slajchert, T., U. Kitron, C. Jones, and A. Mannell. 1997. Role of the eastern chipmunk (*Tamias striatus*) in the epizootiology of Lyme borreliosis in northwestern Illinois, USA. Journal of Wildlife Diseases 33(1):40-46.

12) Schmidt, K., R. Ostfeld, and E. Schauber. 1999. Infestation of *Peromyscus leucopus* and *Tamias striatus* by *Ixodes scapularis* (Acari: Ixodidae) in relation to the abundance of hosts and parasites. Journal of Medical Entomology 36(6): 749-757.

13) Linhart, S., J. Wlodkowski, D. Kavanaugh, L. Motes-Kreimeyer, A. Montoney, R. Chipman, D. Slate, L. Bigler, and M. Fearneyhough. 2002. A new flavor-coated sachet bait for delivering oral rabies vaccine to raccoons and coyotes. Journal of Wildlife Diseases 38(2): 363-377.

14) Lane, R. and J. Loye. 1989. Lyme disease in California: interrelationships of *Ixodes pacificus* (Acari:Ixodidae), the western fence lizard (*Sceloporus occidentalis*) and *Borrelia burgdorferi*. Journal of Medical Entomology 26: 272-278.

15) Rand, P., E. Lacombe, M. Holman, C. Lubelczyk, and R. Smith. 2000. Attempt to control ticks (Acari:Ixodidae) on deer on an isolated island using ivermectin-treated corn. Journal of Medical Entomology 37(1): 126-133

Literature Cited

16) Sonenshine, D. and G. Haines. 1985. A convenient method for controlling populations of the American Dog Tick, *Dermacentor variabilis* (Acari:Ixodidae), in the natural environment. Journal of Medical Entomology 22(5): 577-583.

17) Richter D, A. Spielman, N. Komar, F. Matuschka. 2000. Competence of American robins as reservoir hosts for Lyme disease spirochetes. Emerging Infectious Diseases 6:133-138.

18) LoGiudice, K., R. Ostfeld, K. Schmidt, and F. Keesing. 2003. The ecology of infectious disease: Effects of host diversity and community composition on Lyme disease risk. Proceedings of the National Academy of Sciences, USA. 100(2):567-571.

19) Maxforce® Tick Management System. http://www.maxforcetms.com/

20) The Owl pages: http://www.owlpages.com/owlboxes.html

21) The Raptor Trust: http://www.theraptortrust.org/nestbox.html

22) Welty, J. 1979. The Life of Birds. p96. Saunders College Publishing/Holt Rinehart and Winston. Philadelphia, PA. (Based on an average wt. of 20 grams per deer mouse.)

23) Fipronil http://www.beyondpesticides.org/infoservices/pesticidefactsheets/toxic/fipronil.htm

24) Richardson, J. 2000. Permethrin Spot-On Toxicoses in Cats. The Journal of Veterinary Emergency and Critical Care. 10:103-106.

25) Mutlow, A. and N. Forbes. Haemoproteus in Raptors: Pathogenicity, Treatment, and Control. http://www.lansdown-vets.co.uk/pdf/aavhaemoproteus.pdf

26) World Wildlife Fund. 1999. Hazards and Exposures Associated with DDT and Synthetic Pyrethroids used for Vector Control http://www.worldwildlife.org/toxics/progareas/pop/ddt5.pdf

27) Drummond, R. 2000. Ticks and what you can do about them. p 65-66. Wilderness Press, Berkley, CA.

28) Telephone conversation with Alexander Kovel of EcoHealth, Inc., the manufacturer of Damminix®, 4/7/04.

References for Chapter 5

1) Solberg, V., J. Miller, T. Hadfield, R. Burge, J. Schech, and J. Pound. 2003. Control of *Ixodes scapularis* with topical self-application of permethrin by white-tailed deer inhabiting NASA, Beltsville, Maryland. Journal of Vector Ecology 28: 117-134.

2) Wilson, M., S. Telford III, J. Piesman, and A. Spielman. 1988. Reduced abundance of immature *Ixodes dammini* (Acari: Ixodidae) following elimination of deer. Journal of Medical Entomology. 25:224-228.

3) Deblinger, R., M. Wilson, D. Rimmer, and A. Spielman. 1993. Reduced abundance of immature *Ixodes dammini* (Acari: Ixodidae) following incremental removal of deer. Journal of Medical Entomology. 30: 144-150.

4) Wilson, M., J. Levine, and A. Spielman. 1984. Effect of deer reduction on abundance of the deer tick (*Ixodes dammini*). The Yale Journal of Biology and Medicine. 57:697-705.

5) Stafford, K. C., A. Denicola, and H. Kilpatrick. 2003. Reduced abundance of *Ixodes scapularis* (Acari: Ixodidae) and the Tick Parasitoid *Ixodiphagus hookeri* (Hymenoptera: Encyrtidae) with reduction of white-tailed deer. Journal of Medical Entomology 40:642-652.

6) Rand, P., C. Lubelczyk,, G. Lavigne, S. Elias, M. Holman, E. Lacombe, and R. Smith. 2003. Deer density and the abundance of *Ixodes scapularis* (Acari: Ixodidae). Journal of Medical Entomology 40: 179-184.

7) Daniels, T., D. Fish, and I. Schwartz. 1993. Reduced abundance of *Ixodes scapularis* (Acari:Ixodidae) and Lyme disease risk by deer exclusion. Journal of Medical Entomology 30: 1043-1049.

8) Wilson, M., G. Adler, and A. Spielman. 1985. Correlation between abundance of deer and that of the deer tick, *Ixodes dammini* (Acari: Ixodidae). Annals of the Entomological Society of America 78:172-176.

9) Stafford, K. 1993. Reduced abundance of *Ixodes scapularis* (Acari: Ixodidae) with exclusion of deer by electric fencing. Journal of Medical Entomology. 30(6): 986-996

10) Wilson, M., T. Litwin, T. Gavin, M. Capkanis, D. Maclean, and A. Spielman. 1990. Host-dependent differences in feeding and reproduction of *Ixodes dammini* (Acari: Ixodidae). Journal of Medical Entomology 27: 949-954.

11) Piesman, J. and J. Gray. Lyme Disease/ Lyme Borreliosis. p341. *In* Sonenshine, D. and T. Mather (Eds). 1994. Ecological Dynamics of Tick-Borne Zoonoses. Oxford University Press, New York.

12) Vanderhoof-Forschner, K. 2003. Everything you need to know about Lyme disease and other tick-borne disorders. John Wiley and Sons, Hoboken, N. J. (p65)

13) http://www.dec.state.ny.us/website/dfwmr/wildlife/deer/feedregs.html

Literature Cited

14) http://www.dnr.state.wi.us/org/land/wildlife/hunt/deer/BaitingRegulations.pdf

15) http://www.michigan.gov/documents/supp_fd_facts_83068_7.pdf

16) Sonenshine, D., S. Allan, R. Andrew, I. Norval, and M. Burridge. 1996. A self-medicating applicator for control of ticks on deer. Medical and Veterinary Entomology 10: 149-154.

17) Carroll, J. and M. Kramer. 2003. Winter activity of *Ixodes scapularis* (Acari: Ixodidae) and the operation of deer-targeted control devices in Maryland. Journal of Medical Entomology 40(2): 238-244.

18) Meltzer, M., D. Dennis, and K. Orloski. 1999. The cost effectiveness of vaccinating against Lyme disease. Emerging Infectious Diseases 5(3): 321-328.

19) Dandux outdoors http://www.dandux.com/

References for Chapter 6

1) Mather, T., D. Duffy, S. Campbell. 1993. An unexpected result from burning vegetation to reduce Lyme disease transmission risks. Journal of Medical Entomology 30: 642-645.

2) Wilson, M. 1986. Reduced abundance of adult *Ixodes dammini* (Acari: Ixodidae) following destruction of vegetation. Journal of Economic Entomology 79: 693-696.

3) Ahlgren, C. 1960. Some effects of fire on reproduction and growth of vegetation in northeastern Minnesota. Ecology 41: 431-445.

4) Falco, R. and D. Fish. 1988. Prevalence of *Ixodes dammini* near the homes of Lyme disease patients in Westchester County, New York. American Journal of Epidemiology. 127(4): 826-830.

5) Maupin, G., D. Fish, J. Zultowsky, E. Campos, and J. Piesman. 1991. Landscape ecology of Lyme disease in a residential area of Westchester County, New York. American Journal of Epidemiology 133:1105-1113.

6) Carroll, M., H. Ginsberg, K. Hyland, and R. Hu. 1992. Distribution of *Ixodes dammini* (Acari: Ixodidae) in residential lawns on Prudence Island, Rhode Island. Journal of Medical Entomology 29(6): 1052-1055.

7) Schulze, T., R. Jordan, and R. Hung. 1995. Suppression of subadult *Ixodes scapularis* (Acari:Ixodidae) following removal of leaf litter. Journal of Medical Entomology. 32(5): 730-733.

8) Allan, S. and L. Patrican. 1995. Reduction in *Ixodes scapularis* (Acari: Ixodidae) in woodlots by application of desiccant and insecticidal soap formulations. Journal of Medical Entomology 32(1): 16-20.

9) Patrican, L. and S. Allan. 1995. Application of desiccant and insecticidal soap treatments to control *Ixodes scapularis* (Acari: Ixodidae) nymphs and adults in a hyperendemic woodland site. Journal of Medical Entomology 32(6): 859-863.

10) Curran, K., D. Fish, and J. Piesman. 1993. Reduction of nymphal *Ixodes dammini* (Acari: Ixodidae) in a residential suburban landscape by area application of insecticides. Journal of Medical Entomology 30(1): 107-113.

11) Schultz, T., R. Jordan, L. Vasvary, M. Chomsky, D. Shaw, M. Meddis, R. C. Taylor, and J. Piesman. 1994. Suppression of *Ixodes scapularis* (Acari:Ixodidae) nymphs in a large residential community. Journal of Medical Entomology 31(2): 206-211.

12) Solberg, V., K. Neidhardt, M. Sardelis, F. Hoffman, R. Stevenson, L. Boobar, and H. Harlan. 1992. Field evaluation of two formulations of cyfluthrin for control of *Ixodes dammini* and *Amblyomma americanum*. Journal of Medical Entomology 29(4): 634-638.

13) Schulze, T., G. Taylor, R. Jordan, E. Bosler, and J. Shisler. 1991. Effectiveness of selected granular acaricide formulations in suppressing populations of *Ixodes dammini* (Acari: Ixodidae): short-term control of nymphs and larvae. Journal of Medical Entomology 28(5): 624-629.

14) Schulze, T., W. M. McDevitt, W. Parkin, and J. Shisler. 1987. Effectiveness of two insecticides in controlling *Ixodes dammini* (Acari: Ixodidae) following an outbreak of Lyme disease in New Jersey. Journal of Medical Entomology 24: 420-424.

15) Stafford, K. 1991. Effectiveness of carbaryl applications for the control of *Ixodes dammini* (Acari: Ixodidae) nymphs in an endemic residential area. Journal of Medical Entomology 28(1):32-36.

16) Olkowski, W., H. Olkowski, and S. Daar. 1990. Managing ticks, the least- toxic way. Common Sense Pest Control 6: 4-25.

17) Environmental Protection Agency. 2000. Chlorpyrifos Revised Risk Assessment and Agreement with Registrants. http://www.epa.gov/pesticides/op/chlorpyrifos/agreement.pdf

18) www.beyondpesticides.org/ At this site, search for "ticks".

19) Schmidtmann, E. 1994. Ecologically based strategies for controlling ticks. p240-271, *In* Sonenshine, D. and T. Mather. (Eds.) Ecological Dynamics of Tick-Borne Zoonoses. Oxford University Press, New York.

Literature Cited

20) Lord, R., J. Humphreys, V. Lord, R. McLean, and C. Garland. 1992. *Borrelia burgdorferi* infections in white-footed mice (*Peromyscus leucopus*) in hemlock (*Tsuga canadensis*) habitat in Western Pennsylvania. Journal of Wildlife Diseases 28:364-368.

21) Guerra, M., E. Walker, C. Jones, S. Paskewitz, M. R. Cortinas, A. Stancil, L. Beck, M. Bobo, and U. Kitron. 2002. Predicting the risk of Lyme disease: Habitat suitability for *Ixodes scapularis* in the North Central United States. Emerging Infectious Diseases 8(3):289-297.

22) Schmidtmann, E., J. Schlater, G. Maupin, and J. Mertins. 1998. Vegetational associations of host-seeking adult blacklegged ticks *Ixodes scapularis* Say (Acari: Ixodidae), on dairy farms in Northwestern Wisconsin. Journal of Dairy Science 81: 718-721.

23) Environmental Protection Agency. 2001. Diazinon Revised Risk Assessment and Agreement with Registrants http://www.epa.gov/pesticides/op/diazinon/agreement.pdf

24) LoGiudice, K., R. Ostfeld, K. Schmidt, and F. Keesing. 2003. The ecology of infectious disease: Effects of host diversity and community composition on Lyme disease risk. Proceedings of the National Academy of Sciences, USA. 100(2):567-571.

25) Falco RC, Fish D. 1991. Horizontal movement of adult *Ixodes dammini* (Acari: Ixodidae) attracted to CO_2-baited traps. Journal of Medical Entomology 28: 726-729.

26) Stafford, K. C., A. Denicola, and H. Kilpatrick. 2003. Reduced abundance of *Ixodes scapularis* (Acari: Ixodidae) and the tick parasitoid *Ixodiphagus hookeri* (Hymenoptera: Encyrtidae) with reduction of white-tailed deer. Journal of Medical Entomology 40:642-652.

27) Bloemer, S., G. Mount, T. Arnold Morris, R. Zimmerman, D. Barnard, and E. Snoddy. 1990. Management of Lone Star ticks (Acari:Ixodidae) in recreational areas with acaricide applications, vegetative management, and exclusion of white-tailed deer. Journal of Medical Entomology 27(4): 543-550.

28) Knipling, E. and C. Steelman. 2000. Feasibility of controlling *Ixodes scapularis* ticks (Acari: Ixodidae), the vector of Lyme disease, by parasitoid augmentation. Journal of Medical Entomology 37(5): 645-652.

29) Daniels, T., R. Falco, and D. Fish. 2000. Estimating population size and drag sampling efficiency for the blacklegged tick (Acari:Ixodidae). Journal of Medical Entomology 37(3): 357-363.

30) National Institutes of Health. Household Products Database. http://householdproducts.nlm.nih.gov/index.htm

31) Stafford, K.C. 2001. Tick Control. http://www.caes.state.ct.us/FactSheetFiles/ForestryHorticulture/Tick%20Control01.pdf

32) Environmental Protection Agency. Interim reregistration eligibility decision for carbaryl. http://www.epa.gov/oppsrrd1/REDs/carbaryl_ired.pdf

33) Schulze, T., R. Jordan, R. Hung, R. Taylor, D. Markowski, and M. Chomsky. 2001. Efficacy of granular deltamethrin against *Ixodes scapularis* and *Amblyomma americanum* (Acari: Ixodidae) nymphs. Journal of Medical Entomology. 38(2): 344-346.

References for Chapter 7

1) Gage KL, G. Maupin, J. Montenieri, J. Piesman, M. Dolan, N. Panella. 1997. Flea (Siphonaptera: Ceratophyllidae, Hystrichopsyllidae) and tick (Acarina:Ixodidae) control on wood rats using host-targeted liquid permethrin in bait tubes. Journal of Medical Entomology 34(1):46-51

2) Bloemer, S., E. Snoddy, J. Cooney, and K. Fairbanks. 1986. Influence of deer exclusion on populations of Lone Star ticks and American dog ticks (Acari: Ixodidae). Journal of Medical Entomology 79: 679-683.

3) Burkot, T., G. Mullen, R. Anderson, B. Schneider, C. Happ, and N. Zeidner. 2001. *Borrelia lonestari* DNA in adult *Amblyomma americanum* ticks, Alabama. Emerging Infectious Diseases 7(3): 471-473.

4) Bloemer, S., G. Mount, A. Morris, R. Zimmerman, D. Barnard, and E. Snoddy. 1990. Management of Lone Star ticks (Acari: Ixodidae) in recreational areas with acaricide applications, vegetation managements, and exclusion of white-tailed deer. Journal of Medical Entomology 27(4): 543-550.

5) Lane, R.S., J. Piesman, and W. Burgdorfer. 1991. Lyme borreliosis: Relation of its causative agent to its vectors and hosts in North America and Europe. Annual Review of Entomology. 36: 587-609.

6) Kuo, M., R. Lane, and P. Giclas. 2000. A comparative study of mammalian and reptilian alternative pathway of complement-mediated killing of the Lyme disease spirochete (*Borrelia burgdorferi*). Journal of Parasitology 86(6): 1223-1228.

7) Lane, R., and G. Quistad. 1998. Borreliacidal factor in the blood of the western fence lizard (*Sceloporus occidentalis*). Journal of Parasitology 84: 29-34.

8) Casher, L., R. Lane, R. Barrett, and L. Eisen. 2002. Relative importance of lizards and mammals as hosts for ixodid ticks in northern California. Experimental and Applied Acarology 26: 127-143.

9) Centers for Disease Control. Southern Tick-Associated Rash Illness. http://www.cdc.gov/ncidod/dvbid/stari/

Literature Cited

10) Drummond, R. 2000. Ticks and what you can do about them. p 16-17. Wilderness Press, Berkley, CA.

11) Magnarelli, L., J. Anderson, and D. Fish. 1987. Transovarial transmission of *Borrelia burgdorferi* in *Ixodes dammini* (Acari: Ixodidae). The Journal of Infectious Diseases 156(1): 234-236.

12) Oliver, J. 1996. Lyme borreliosis in the Southern United States: A review. Journal of Parasitology 82(6): 926-935.

13) Li, X., C. Peavy, and R. Lane. 2000. Density and spatial distribution of *Ixodes pacificus* (Acari: Ixodidae) in two recreational areas in north coastal California. American Journal of Tropical Medicine and Hygiene 62(3): 415-422.

14) Talleklint-Eisen, L. and R. Lane. 1999. Variation in the density of questing *Ixodes pacificus* (Acari: Ixodidae) nymphs infected with *Borrelia burgdorferi* at different spatial scales in California. Journal of Parasitology 85(5): 824-831.

15) Brown, R. and R. Lane. 1992. Lyme disease in California: A novel enzootic transmission cycle of *Borrelia burgdorferi*. Science 256: 1439-1442.

16) Layfield, D. and P. Guilfoile. 2002. The prevalence of *Borrelia burgdorferi* (Spirochaetales: Spirochaetaceae) and the Agent of Human Granulocytic Ehrlichiosis (Rickettsiaceae: Ehrlichieae) in *Ixodes scapularis* (Acari:Ixodidae) Collected During 1998 and 1999 from Minnesota. Journal of Medical Entomology 39(1): 218–220.

17) Vanderhoof-Forschner, K. 2003. Everything you need to know about Lyme disease and other tick-borne disorders. John Wiley and Sons, Hoboken, N. J.

18) Dr. Wayne Rowley, Iowa State University, personal communication.

19) Varela, A., M. Luttrell, E. Howerth, V. Moore, W. Davidson, D. Stallknecht, and S. Little. 2004. First culture isolation of *Borrelia lonestari*, putative agent of Southern Tick-Associated Rash Illness. Journal of Clinical Microbiology 42(3): 1163-1169.

Glossary

Blacklegged ticks; small ticks capable of transmitting Lyme disease. Found in most of the eastern half of the United States, their scientific name is *Ixodes scapularis.*

Competence; As used in this book, the ability of a host to support the growth of a pathogen and transmit the pathogen to a tick.

Host; an animal that acts as a source of a blood meal for a tick.

Larvae; the life stage of a tick that hatches from an egg. Blacklegged tick larvae are about the size of a poppy seed and typically feed on mice. If they successfully feed and molt, they will become nymphs. See Figure 2.1 on page 17.

Lone Star tick; a relatively large tick, found in the Southeastern and mid-central United States. This tick appears to transmit a pathogen that causes a disease similar to Lyme disease. The scientific name of this tick is *Amblyomma americanum.*

Lyme bacterium; The pathogen that causes Lyme disease. The scientific name for this organism is *Borrelia burgdorferi.*

Glossary

Nymph; an intermediate stage in the life cycle of ticks. Blacklegged tick nymphs are about the size of a sesame seed, and are most likely to transmit Lyme disease since victims often don't see these small ticks biting them. Nymphs, if they feed successfully and molt, become adults. See Figure 2.1 on page 17.

Vector; an organism, such as a tick, capable of transmitting disease from one host to another.

Western blacklegged tick; a tick, closely related to blacklegged ticks, found along the Pacific Coast and part of the Southwest United States. This organism can transmit the Lyme bacterium. Its scientific name is *Ixodes pacificus.*

Index

B

birds 19, 35, 37, 53
blacklegged ticks 5, 7, 17, 18, 20, 21, 23, 24, 25, 27, 31, 34, 35, 37, 38, 39, 47, 48, 49, 51, 52, 55, 56, 57, 60, 61, 62, 73, 77
burning 45, 46, 71

C

Carbaryl 51
cyfluthrin 51, 72

D

Damminix 32
deer mice 19, 20, 22, 29, 30, 31, 33, 38, 53, 56, 57, 61
DEET 12

F

fence 40, 68, 74
flagging 25, 49, 67
4-poster 8, 41, 43
fungi 52

I

Ixodes pacificus 7, 67, 68, 75, 77
Ixodes scapularis 7, 17, 27, 63, 64, 65, 66, 67, 68, 69, 70, 71, 72, 73, 75, 76

L

lawns 46, 47, 53, 62, 71
leaf litter 18, 24, 46, 47, 48, 53, 54, 62, 71
leaves 45, 46, 48, 49, 54
lizards 19, 37, 56, 57, 74
Lone Star ticks 7, 16, 58, 60, 73, 74
Lyme
    bacterium 6, 8, 5, 6, 7, 9, 11, 13, 14, 15, 17, 20, 21, 22, 23, 24, 26, 30, 31, 32, 34, 35, 36, 37, 38, 39, 40, 41, 42, 43, 46, 53, 55, 56, 57, 58, 63, 64, 65, 66, 67, 68, 69, 70, 71, 72, 73, 74, 75, 76, 77

disease 6, 8, 5, 6, 7, 9, 11, 13, 14, 15, 17, 20, 21, 22, 23, 24, 26, 30, 31, 32, 34, 35, 36, 37, 38, 39, 40, 41, 42, 43, 46, 53, 55, 56, 57, 58, 63, 64, 65, 66, 67, 68, 69, 70, 71, 72, 73, 74, 75, 76, 77

M

Midwest 9, 19, 20, 35, 55, 57
mowing 45, 46, 62

N

Northeastern U.S. 7, 9, 20, 26, 55, 57
nymph 18, 22, 50

P

Pacific Coast 26, 55, 56, 77
permethrin 13, 33, 41, 50, 64, 67, 68, 69, 74
pesticide 32, 33, 34, 38, 41, 43, 44, 50, 51, 52, 62

S

Southeastern U.S. 26, 55, 57, 58
squirrels 36, 53, 62

T

Tick. *See also* blacklegged ticks
  flag 24, 25
  traps 25, 49, 60, 67, 73

V

vaccine 14, 37, 63, 64, 68

W

wasps 52
western blacklegged ticks 55, 56, 57
white-tailed deer 18, 19, 21, 29, 39, 57, 69, 70, 73, 74

Ticks Off! Order form

Name _____

Address: _____

_____

City _____

State _____

Zip _____

Please send me the following number of copies of "Ticks Off! Controlling Ticks That Transmit Lyme Disease on Your Property"

# of books requested _____

X $9.95 per book _____

MN residents add 6.5% sales tax ($0.65 per book) _____

Total _____

Free shipping in the continental U.S!

Send check or money order to:
ForSte Press, Inc
P.O. Box 1537
Bemidji, MN 56619-1537

Or order on the Web: www.forstepress.com

Questions? Contact us at: books@forstepress.com

Visit www.tickbook.com for a complete list of web addresses from the book, updates, and other information relevant to controlling ticks.